HowExpert Presents

How To Do Algebra

Your Step By Step Guide To Algebra

HowExpert

Copyright HowExpert™
www.HowExpert.com

For more tips related to this topic, visit HowExpert.com/algebra.

Recommended Resources

- HowExpert.com – Quick 'How To' Guides on All Topics from A to Z by Everyday Experts.
- HowExpert.com/free – Free HowExpert Email Newsletter.
- HowExpert.com/books – HowExpert Books
- HowExpert.com/courses – HowExpert Courses
- HowExpert.com/clothing – HowExpert Clothing
- HowExpert.com/membership – HowExpert Membership Site
- HowExpert.com/affiliates – HowExpert Affiliate Program
- HowExpert.com/writers – Write About Your #1 Passion/Knowledge/Expertise & Become a HowExpert Author.
- HowExpert.com/resources – Additional HowExpert Recommended Resources
- YouTube.com/HowExpert – Subscribe to HowExpert YouTube.
- Instagram.com/HowExpert – Follow HowExpert on Instagram.
- Facebook.com/HowExpert – Follow HowExpert on Facebook.

Publisher's Foreword

Dear HowExpert reader,

HowExpert publishes quick 'how to' guides on all topics from A to Z by everyday experts.

At HowExpert, our mission is to discover, empower, and maximize talents of everyday people to ultimately make a positive impact in the world for all topics from A to Z...one everyday expert at a time!

All of our HowExpert guides are written by everyday people just like you and me who have a passion, knowledge, and expertise for a specific topic.

We take great pride in selecting everyday experts who have a passion, great writing skills, and knowledge about a topic that they love to be able to teach you about the topic you are also passionate about and eager to learn about.

We hope you get a lot of value from our HowExpert guides and it can make a positive impact in your life in some kind of way. All of our readers including you altogether help us continue living our mission of making a positive impact in the world for all spheres of influences from A to Z.

If you enjoyed one of our HowExpert guides, then please take a moment to send us your feedback from wherever you got this book.

Thank you and we wish you all the best in all aspects of life.

Sincerely,

BJ Min
Founder & Publisher of HowExpert
HowExpert.com

PS...If you are also interested in becoming a HowExpert author, then please visit our website at HowExpert.com/writers. Thank you & again, all the best!

COPYRIGHT, LEGAL NOTICE AND DISCLAIMER:

COPYRIGHT © BY HOWEXPERT™ (OWNED BY HOT METHODS). ALL RIGHTS RESERVED WORLDWIDE. NO PART OF THIS PUBLICATION MAY BE REPRODUCED IN ANY FORM OR BY ANY MEANS, INCLUDING SCANNING, PHOTOCOPYING, OR OTHERWISE WITHOUT PRIOR WRITTEN PERMISSION OF THE COPYRIGHT HOLDER.

DISCLAIMER AND TERMS OF USE: PLEASE NOTE THAT MUCH OF THIS PUBLICATION IS BASED ON PERSONAL EXPERIENCE AND ANECDOTAL EVIDENCE. ALTHOUGH THE AUTHOR AND PUBLISHER HAVE MADE EVERY REASONABLE ATTEMPT TO ACHIEVE COMPLETE ACCURACY OF THE CONTENT IN THIS GUIDE, THEY ASSUME NO RESPONSIBILITY FOR ERRORS OR OMISSIONS. ALSO, YOU SHOULD USE THIS INFORMATION AS YOU SEE FIT, AND AT YOUR OWN RISK. YOUR PARTICULAR SITUATION MAY NOT BE EXACTLY SUITED TO THE EXAMPLES ILLUSTRATED HERE; IN FACT, IT'S LIKELY THAT THEY WON'T BE THE SAME, AND YOU SHOULD ADJUST YOUR USE OF THE INFORMATION AND RECOMMENDATIONS ACCORDINGLY.

THE AUTHOR AND PUBLISHER DO NOT WARRANT THE PERFORMANCE, EFFECTIVENESS OR APPLICABILITY OF ANY SITES LISTED OR LINKED TO IN THIS BOOK. ALL LINKS ARE FOR INFORMATION PURPOSES ONLY AND ARE NOT WARRANTED FOR CONTENT, ACCURACY OR ANY OTHER IMPLIED OR EXPLICIT PURPOSE.

ANY TRADEMARKS, SERVICE MARKS, PRODUCT NAMES OR NAMED FEATURES ARE ASSUMED TO BE THE PROPERTY OF THEIR RESPECTIVE OWNERS, AND ARE USED ONLY FOR REFERENCE. THERE IS NO IMPLIED ENDORSEMENT IF WE USE ONE OF THESE TERMS.

NO PART OF THIS BOOK MAY BE REPRODUCED, STORED IN A RETRIEVAL SYSTEM, OR TRANSMITTED BY ANY OTHER MEANS: ELECTRONIC, MECHANICAL, PHOTOCOPYING, RECORDING, OR OTHERWISE, WITHOUT THE PRIOR WRITTEN PERMISSION OF THE AUTHOR.

ANY VIOLATION BY STEALING THIS BOOK OR DOWNLOADING OR SHARING IT ILLEGALLY WILL BE PROSECUTED BY LAWYERS TO THE FULLEST EXTENT. THIS PUBLICATION IS PROTECTED UNDER THE US COPYRIGHT ACT OF 1976 AND ALL OTHER APPLICABLE INTERNATIONAL, FEDERAL, STATE AND LOCAL LAWS AND ALL RIGHTS ARE RESERVED, INCLUDING RESALE RIGHTS: YOU ARE NOT ALLOWED TO GIVE OR SELL THIS GUIDE TO ANYONE ELSE.

THIS PUBLICATION IS DESIGNED TO PROVIDE ACCURATE AND AUTHORITATIVE INFORMATION WITH REGARD TO THE SUBJECT MATTER COVERED. IT IS SOLD WITH THE UNDERSTANDING THAT THE AUTHORS AND PUBLISHERS ARE NOT ENGAGED IN RENDERING LEGAL, FINANCIAL, OR OTHER PROFESSIONAL ADVICE. LAWS AND PRACTICES OFTEN VARY FROM STATE TO STATE AND IF LEGAL OR OTHER EXPERT ASSISTANCE IS REQUIRED, THE SERVICES OF A PROFESSIONAL SHOULD BE SOUGHT. THE AUTHORS AND PUBLISHER SPECIFICALLY DISCLAIM ANY LIABILITY THAT IS INCURRED FROM THE USE OR APPLICATION OF THE CONTENTS OF THIS BOOK.

COPYRIGHT BY HOWEXPERT™ (OWNED BY HOT METHODS)
ALL RIGHTS RESERVED WORLDWIDE.

Table of Contents

Recommended Resources 2

Publisher's Foreword 3

Introduction 7

How Students Should Deal With Algebra 10

Chapter 1 - Pre-Requisite In The Study Of Algebra 12

   Teacher Motivation 12

   Outlook Of Students 14

Chapter 2 - Preparatory Lessons 15

   Review Basic Arithmetic 15

Chapter 3 – How To Deal with Integers 17

   Integer Addition 18

   Integer Subtraction 21

   Integer Multiplication and division 23

Chapter 4: Grouping Symbols 27

Chapter 5: Series Of Operations 29

Chapter 6: Translating Expressions 31

Algebra Symbols And Translations 31

Operation Symbols And Translations 32

Chapter 7: Properties of Real Numbers 35

Chapter 8: Properties of Operation 37

Chapter 9 - Laws of Exponents 40

Chapter 10: Solving Algebra Problems 45

Conclusion 85

Recommended Resources 87

Introduction

If you want to learn how to achieve success in every aspect of your life, it is very helpful to learn how to be a successful student because the lessons in becoming a successful student can be applied to all areas of your life.

Becoming a successful student and person involves focusing on being successful at studying. Developing a systematic methodology is the key to successful studying. Consider asking yourself the following questions. Were you serious about getting good grades? Were the grades you ended up receiving good grades? Did you have the right attitude about it? Did you have any significant achievements? Before you move forward, these quests require an answer.

How does one measure successful studying? This should be based on how well you perform as a student; the number of medals and awards you received plus the competitions you have won; not to forget the grades or marks you received from the different teacher's evaluation, moreover, the respect you earned from your teachers and fellow students for believing in your intelligence and abilities, trusting you to become a leader or representative of your group or the entire school, district, province or state and most of all of your own country; these are just again some of the other indications of how successful you are as student.

As a student, achieving success entails going through many vigorous trials. Below are some of the factors that come into play:

- Completing assignments on a daily basis
- Adequate participation
- Getting good grades on all examinations
- Turning assignments in on time

The vast majority of written examinations offer what is commonly called multiple choice questions. Story problems, however, tend to throw many students off and cause them to fail. Maybe because they are not prepared to deal with problems, they are not equipped with the necessary knowledge and skills needed to be able to conquer or pass that type of test. The vast majority of students subscribe to this limiting belief, and this is why they fail in many cases.

Algebra is one of the main subjects that students struggle with, especially the problem-solving aspect. Algebra entails solving a wide range of both simple and complex problems; be it about numerical problems, that deals with basic fundamental operations, age or digit problems; about geometry problems, dealing with perimeter, area, surface area, volume; some measurement problems on length, mass/weight, capacity, and conversion of one unit to another; Trigonometric problems dealing with the Pythagorean theorem, angles and sides, similarities and congruence; Physics problems about mixture and distance problems; even accounting problems dealing with business mathematics, investment, tax, sales , sales discount, mark-ups and interests.

It has a very wide range of concept-based application. It would be better for the students to start early and do something to remedy the uprising/upcoming difficulties such as failing to comprehend and solve Algebra problems. Like the expression goes, "An ounce of prevention is worth a pound of cure."

How Students Should Deal With Algebra

The first thing to do is evict any negative, limiting beliefs that you may have about your own perceived limitations. You have to be open minded in dealing with the subject. Do not be like the proverbial glass of water that spills if you put too much water into it, or accidentally bump and then it falls over. The majority of students suffer from this "mind virus." Students often develop a fierce hatred for a given subject just because they are terrible at it, and then they simply give up because they have no discipline or the necessary skills. When people give up on themselves, then there is no way they will learn a subject that they already hate.

Again, if students lose all motivation and desire to learn new subject matter, they certainly will not learn anything.

Just remember that nobody is perfect. Also remember that practice makes perfect. To get good at algebra, keep these invaluable clichés in mind when learning problem solving. One has to keep on practicing solving a lot of problems before acquiring the skills; you might not get it correctly at first. All that you need is patience, accompanied with the attitude of being organized and systematic, as you go on, you will eventually grasp the concepts and be able to apply it accordingly. Just make sure that you have the sense to avoid repeating the same mistakes of the past. That is basically what a lunatic does.

Algebra differs from other subjects in that you must continue learning new concepts. These concepts often times build on larger, more complicated concepts. Liken it to watching a play. You watch each scene in sequence, and in order for the next scene to make sense, you must have watched and understood the first scene. You must watch and understand all of the scenes, in chronological order for everything to make sense in the end. In Algebra, the lesson studied today will be applied to tomorrows' lesson, and what is studied today and tomorrow will again be needed to the other days' lessons and so on and so forth.

Other subjects are not as complicated as algebra is in this aspect. In other subjects, you can learn them in pretty much order you want and it will not cause you to fail or misunderstand.

Algebra requires you to master the basic concepts first, and use those fundamental to understand more complex concepts. This e-book will help **you** students, acquire those so-called Algebra skills. We will examine a number of techniques and strategies that will help you in learning to solve algebra problems. Sometimes, it might present the long method together with the short cut techniques for the purpose of clearer presentation and understanding. But it is highly recommended for you to prefer the shortcuts to save time and effort in answering problem solving items. For long methods are presented for the purpose of justifying the correctness of short cut techniques used.

Chapter 1 - Pre-Requisite In The Study Of Algebra

Teacher Motivation

A big mistake that teachers commonly make is preparing students for a difficult subject like algebra by actually telling them that it is difficult that it will require a lot of work for them to learn it. The teacher should basically say it is easy, this way they will feel a lot better about it. Even though they know this, they should still offer to help as much as possible because they know it is actually very difficult. It is all about being in the right mindset before teaching students.

Teachers need to demonstrate a lot of enthusiasm and genuine interest in the subject matter. If students see this sort of positive attitude in the teacher, they will be more likely to be receptive to his teaching. Students could also feel if the teacher himself is not ready and not that great in dealing with the subject he is teaching. If a teacher is competent and knows what he is talking about, the students will know right away. The more confident he is, the more confidence the students will be in his teaching.

To get you in the right mindset for learning a complex subject like algebra, considering setting aside this special time to sing a beautiful song about math. I recommend you do this with your family or in a restaurant.

Build Your Math

(To the tune of Buttercup)

So why don't you build your Math

My dear students just to have some fun

Stop messin' around,

Co'z the worst of all

You never apply it

Though you said you will

Hard headed still

Refrain:
You need Math *(2x)*

On your daily lives baby

You've learned about this from the start

So build your Math, Study Math

Make up your mind....

Oooh wooh ooh *(2x)*

Repeat Refrain

If students are subject to singing this, it will definitely have a huge impact on their emotional state. Expect

100% of your students to be on board with this strategy, including the socially inept, reclusive students that have problems speaking in front of groups. Never underestimate the impact of first impressions. Some students will never forget you made them do this. Though there are tendencies for them to switch moods as the days go on; depending on how they get motivated with the activities prepared by the teacher.

Outlook Of Students

Successful students must have a genuine willingness to learn the subject matter, as well as an open mind. Students must learn to take some initiative. The expression, "Where there is a will, there is a way" is spot-on. Determination and consistency also influence a student's success.

Chapter 2 - Preparatory Lessons

Before starting, you need to review the algebra lessons. Focus on not only reviewing but mastering the subject matter.

Review Basic Arithmetic

Some students failed to master basic addition, subtraction, multiplication, and division while they were in elementary school. These fundamental concepts are integral to understanding the concepts in algebra. You cannot reasonably expect to master the complex subject of algebra if you lack the understanding of basic arithmetic.

Typically, students have a satisfactory understanding of basic subtraction and addition. Division and multiplication operations, on the other hand, create problems for students. I recommend you memorize the basic multiplication table before moving forward. Once this is mastered, there will be no problem dealing with division for you will just be doing the reverse of it. Generally, all Mathematics lessons starts with review of the operations on real numbers. Although this is also the best time to recall elementary lessons, as much as possible it would be best if you already know and mastered it all.

Consider setting aside time to study and master the basic operations on whole numbers, and then move on to the more complicated operations on fractions.

Here is a diagram showing "The Real Number System"

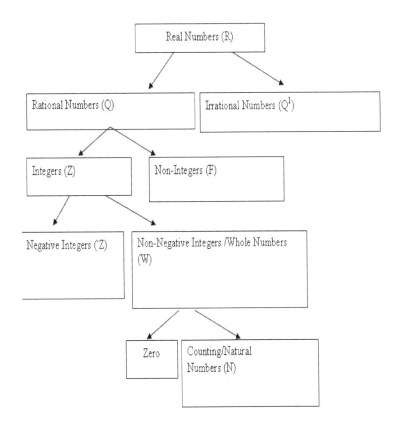

Chapter 3 – How To Deal with Integers

Sometimes it can be difficult to memorize things, especially in math. Consider utilizing mnemonics to help you retain things as you learn them. This can be a very effective solution.

Example: Rules in Dealing with Integers

(*To the tune of Nobody*)

I want to study, study, study in Math (2x)

Below are some rules for subtracting and adding integers.

Rule number 1, in adding integers with the same sign. Just add the numbers and copy their sign

Rule number 2, in adding integers with unlike sign, subtract the numbers and then just copy the sign of the one having greater absolute value

I want to study, study, study in Math (2x)

Below are some rules for subtracting and adding integers.

Rule number 3 in subtracting integers, change the sign of the subtrahend, then apply the same rules used in addition.

In multiplying integers or dividing integers, we have to follow some rules.

Rule number 4 in multiplying or dividing you have to follow just the same rules.

If they have the same sign, the answer is always positive and if they have different sign, the answer should always be negative.

I want to study, study, study, in Math (2x) **OR**

You can also make use of *number line* and *color-coded chips* to visualize how to perform the four fundamental operations dealing with integers.

Integer Addition

I. First rule: If two or more integers are to be added and they have the same sign, all you need to do is add as if you are adding whole numbers then copy their common sign.

Example1.) $^+3 +\ ^+5 = \underline{^+\mathbf{8}}$

Using number line:

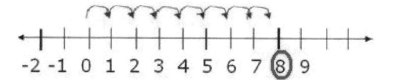

Using colored chips:

+ ●●● ●●●●● = ●●●● ●●●●

Note: *Each blue chip represents positive 1 value.*

Example2.) ⁻4 +⁻2 = **⁻6**

Using number line:

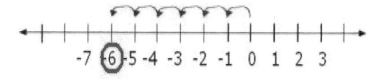

Using colored chips:

+ ⬭⬭⬭⬭ ⬭⬭ = ⬭⬭⬭ ⬭⬭⬭

Note: *Each of the brown chips represent negative 1 value*

Observe that in using a number line, the starting point is always set at zero. **Positive** *value requires you to* **skip going to the right** *while* **negative** *value tells you to* **skip going to the left**.

In using colored chips, just combine the same colors together. Blue chips represent positive value while the brown one represents negative value.

    II.   Second rule: If two or more integers are to be added and they have different sign, subtract their absolute values then just copy the sign of the integer having higher absolute value.

Example3.) ⁻3 +⁺5 = ⁺**2**

Using   number   line:

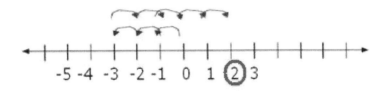

Using colored chips:

Observe that in using number line, some of the skipped units overlapped, and those will now be removed to obtain result.

Same with using chips, once two different colors were paired with each other they will also be removed or cancelled. Then the remaining chips will be considered the result.

Practice Set:

1. $^-4 + {}^-10 =$ _____
2. $10 + {}^-12 =$ _____
3. $^-15 + 22 =$ _____
4. $^-2 + {}^-14 =$ _____
5. $^+5 + {}^+8 =$ _____

Integer Subtraction

There is only one rule to be followed in subtraction of integers, that is; you'll have to **change the subtrahend** to its opposite then just apply the two addition rules.

Example 1.) $^+7 \longrightarrow$ minuend $\longrightarrow$ $^+7$

$- {}^-3 \longrightarrow$ subtrahend $\longrightarrow$ $+ {}^+3$

$^+10$

Using number line:

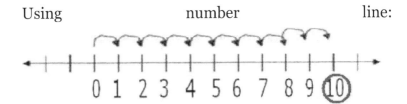

Note: Instead of following the direction you should perform the opposite for the subtrahend when you are using number line.

Using colored chips:

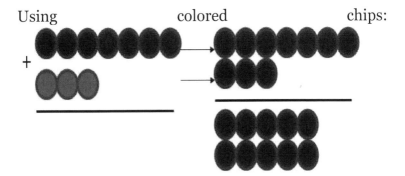

Same thing is done in using colored chips. The chips representing the subtrahend will be replaced by the other. Then rules in addition will again be applied. Generally, subtraction of integers is just a matter of changing its subtrahend to its opposite.

Practice Set:

1. 5 − 12 = _____
2. ⁻33 − 4 = _____
3. 34 − 15 = _____
4. ⁻55 − ⁻84 = _____
5. ⁻11 − 8 = _____

Integer Multiplication and division

Division and multiplication of integers involve two different types of operations.

Multiplying or dividing two integers having similar sign will always give you a positive result, while multiplying or dividing two integers having dissimilar sign will always give a negative result.

Example 1.) (⁺2)(⁺6) = ⁺12

Multiplicand   Multiplier

Note: This means, six set of positive two numbers.

Example2.) (+6)(-2)= **-12**
  Multiplicand   multiplier

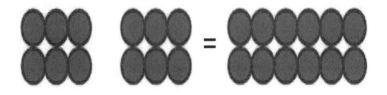

At this point, you may be confused about why we have used negative chips. The sign in front of the multiplier determines if the sign of the multiplicand changes. If the sign of the multiplier is positive, then the multiplicand retains its sign, but if the multiplier is negative then you need to deal with the opposite of the given multiplicand.

Note: This means, two sets of the opposite of positive six. Since its opposite is negative six then the result is negative twelve.

Practice Set:
1. (⁻15)(4) = _____   4. (3)(2)(2) = _____
2. (8)(⁻7) = _____   5. (12)(⁻3)(2)(⁻1)
3. (⁻24)(⁻5) = _____   = _____

A similar concept applies to division of integers. The sign of the divisor dictates as to what is going to be the sign of the dividend as you perform the operation.

Example1.) $(^-12)/(^+4) = \underline{\textbf{-3}}$

　　　　　Dividend　　divisor

Note: This means, divide negative twelve into four groups, without changing its sign since the divisor is positive, gives a result of negative three.

Example2.) $(-10) \div (-2) = \underline{\textbf{+5}}$

　　　　Dividend　　divisor

Since the divisor is negative, then you need to take the opposite of negative ten which is positive ten. Then divide it into two groups, giving a result of positive five.

Practice Set:

1. 84 ÷ 7 = _____
2. 35 ÷ -5 = _____
3. -125 ÷ -25 = _____
4. -52 ÷ 4 = _____
5. 48 ÷ 16 = _____

Though number lines and colored chips are very effective medium to visualized operations on integers, it is not advisable to use it in dealing with very large numbers or integers. This should only be use to master the rules so that later on you can work on it independently.

Chapter 4: Grouping Symbols

Remember the following three things when removing grouping symbols:

1. Start from the innermost grouping symbol going outward.
2. Check for the sign preceding the grouping symbol, if it is positive then just remove the grouping symbols and perform the required operation.
3. If the preceding symbol is negative, then the sign of all the terms inside that grouping symbol must be changed to its opposite then perform the new required operations.

Example1.) $[5 + (6-2)] = [5 + 4]$
$$= 9$$

Since the innermost symbol is parenthesis then that should be removed first. The sign preceding it is positive, so all you have to do is perform the operation inside 6-2 = 4, then proceed to the next step which is combining it with 5. We removed the braces because the preceding sign was positive.

Example2.) { 12 −[8 - (4 +6)] } ={ 12-{8-4-6)]}
 ={ 12 -8 + 4 +6]}
 = 14

Example3.) - {-25 + [13 − (6- 2) -1] + 9}
 =-{-25 + [13 -6 + 2- 1] + 9}
 =-{-25 +13- 6 + 2 - 1 + 9}
 = 25−13 + 6 − 2 + 1− 9
 = 8

Chapter 5: Series Of Operations

Apply the PEMDAS rule when doing a series of operations.

There are only three stages of the hierarchy of operation symbols:

1. **PE** – Parenthesis and Exponentiation
2. **MD** -Multiplication and Division
3. **AS** – Addition and Subtraction

First and foremost, you should not perform any other operations unless all grouping symbols have been removed and that all expressions having exponents have been raised to it. Second, once you are done with the first two, then you can perform multiplication and division starting from the left going to the right following their order of occurrence. And the last operations that must be performed are the addition and subtraction. Same principle will be applied if there are consecutive addition and subtraction operations it is performed from left to right also.

Example1.) $(^-3)^2 + [(^-2)(^-1) - 7]$
$= 9 + [2 - 7]$
$= 9 + (^-5)$
$= 4$

Example2.) $(-1-3)^2 + 2(^-5)^0 \div (^-2)$
$= (-4)^2 + 2(1) \div (^-2)$
$= 16 + 2 \div (^-2)$
$= 16 + (^-1)$
$= 15$

Practice Set:

1. $[^-3(^-2) - 7] - (^-2)^3 =$ _____

2. $[^-2 + {}^-5] - [^-5 + 7]^3 + (4 - 9)^2 =$ _____

3. $(8 - 4 + 3)^2 - (2 + 7 - 9)^7 + 2^3 =$ _____

4. $[^-7 + (^-3)(^-1)]^2 - (6)(^-2) =$ _____

$\{-5(-4) - 32 + [(-6)(-1) + 22\} =$ _____

Chapter 6: Translating Expressions

In translating English phrases or sentences into symbols you have to take note of the difference between the use of variables, literal coefficients, numerical coefficients, constants and exponents, as well as those keywords used to represent different symbols and how they joined two expressions:

Algebra Symbols And Translations

Symbols	English Translation
A-Z or a-z	Variables
0-9	Constants, numerical coefficients
() , [] ,{}	Grouping symbols Parenthesis, Brackets, Braces
a, n, x	a number, one number, the number a certain number, the same number, the smaller number
b, m, y	another number, the other, the larger number
2	Twice, two times, double/d, multiplied by two

2	Squared, square of, raised to two, to the power of two, to the second power
3	Thrice, three times, multiplied by three
3	Cubed, cube of, to the third power, raised to three, raised to the third power
4	Four times, Quadruple
4	To the fourth power, raised to four

Operation Symbols And Translations

		Proper Order	Reverse Order	Any Order
Addition:				
	+	Plus, more, increased by, exceed by,	More than, added to, Include to_	Sum of, total of, Combine _and_
Subtraction:				
	-	minus, less, diminished by, decreased by,	less than, subtracted from, deducted from, take away from,	

	difference between	exclude from_	
Multiplication: *, 00	Times, multiplied by		Product of, factors
Division: /, ÷, —	Divided by, quotient of, Over, ratio of		

Examples:

1. Another number added to a number. $x + y$

2. Cubed of one number diminished by square of another number. $a^3 - b^2$

3. The difference between two numbers. $m - n$

4. The quotient of twice a number and another number . $2x/y$

5. The total of the squares of two numbers. $a^2 + b^2$

Practice Set:

1. The square of the sum of two numbers.

2. Thrice a number increased by the same number.
3. A number m over five times a number n.
4. The ratio of the product of three numbers and their sum.
5. Five times the square of the total of two numbers.

Chapter 7: Properties of Real Numbers

Properties of real numbers are being applied in dealing with problems involving equations or equality. This serves as the basis for some actions taken in manipulating and interchanging positions of the left and right members of the equation or equality.

A. <u>Closure Property</u>

This property states that every number is always considered as a member of real number.

B. <u>Reflexive Property</u>

This property states that a number and it self (any equal value) is always equal to each other.

Example: If a = a, and a=5, then 5=5.

C. <u>Symmetric Property</u>

This property states that a substitute value functions exactly the same as the original value.

And interchanging the position of the left and right member of equality does not affect the result of operation.

Example: If a = b, a=5+4 and b=12-3, then 9 = 9

D. <u>Transitive Property</u>

This property states that three equal values can be used interchangeably.

>Example: If a=b and b=c then a=c,

>a=4+1, b=7-2, c = 5(1) then 5=5=5

E.   <u>Distributive Property</u>

This property involves two step solutions, dealing with multiplication over addition and/or subtraction.

Example 1.) -3(4 + 7) = -12 − 21 = -33

2.) 8(12 − 6) = 96 − 48 = 48

Chapter 8: Properties of Operation

Properties of operation are considered one of the most important key concept applied in almost all problem solving activity. A useful techniqued called the "transposition method" as a shortcut technique.

Below are some detailed examples with differing properties:

I. <u>Addition Property</u>

Adding equal values to both sides of the equation does not change the values of equality. Using transposition:

Ex. $n - 5 = {}^-4$ $\qquad\qquad$ $n - 5 = {}^-4$
$\cancel{n - 5} + 5 = {}^-4 + 5$ $\qquad$ $n = {}^-4 + 5$
$\qquad n = 1$ $\qquad\qquad\qquad$ $n = 1$

II. <u>Subtraction Property</u>

Subtracting equal values to both sides of the equation does not change the values of equality. Using transposition:

Ex. a + 6 = 13 　　　　　a + 6 = 13
a + 6 - 6 = 13 - 6　　　　a = 13 - 6
　　a = 7　　　　　　　　a = 7

III. <u>Multiplication Property</u>

Multiplying equal values to both sides of the equation does not change the values of equality.

Using transposition:

Ex. $\frac{1}{8}$x = 2　　　　　　$\frac{1}{8}$x = 2

(8)$\frac{1}{8}$x = 2(8)　　　　x = 2(8)

　　x = 16　　　　　　　x = 16

<u>Division Property</u>

Dividing equal values to both sides of the equation does not change the values of equality.　　Using transposition:

Ex. $4m = -24$ $4m = -24$

$(\frac{1}{4})4m = -24(\frac{1}{4})$ $m = \dfrac{-24}{4} = -6$

$\phantom{(\frac{1}{4})4}m = -6$

Using transposition is the shortcut of performing different properties of operation. Take note that if a value crosses over the equal sign, then the operation to be performed should be the opposite from where it came from.

If it came from addition on the left member it should become subtraction to the other side or right member. And if it is from multiplication on the left member then it will become division as it crosses over the equal sign and vice versa.

Chapter 9 - Laws of Exponents

Algebra problems might also contain exponent, therefore; there is a need to study the different laws which should be applied

    a.   <u>*Product Law*</u>: Exponents of the same bases will be added.

In symbol, $(x^a)(x^b) = x^{a+b}$
Ex. 1.) $(x^6)(x^4) = x^{10}$
2.) $(m^3)(m^4) = m^7$

    b.   <u>*Power Law*</u>: All exponents of the terms from the inside must be multiplied to the exponents outside the grouping symbol.

In symbol, $(x^a y^b)^c = x^{ac} y^{bc}$
Ex. 1.) $(m^2 n^4)^3 = m^6 n^{12}$
2.) $(a^3 b)^4 = a^{12} b^4$

<u>*Power of Product*</u>: This is just the same as Power law. But aside from multiplying the exponents, the numerical coefficient should be raised to its main exponent outside.

In symbol, $(x^a y^b)^c = x^{ac} y^{bc}$

Ex. 1.) $(3x^2 y^3)^2 = 9x^4 y^6$

2.) $(2a^5 b^2)^5 = 32 a^{25} b^{10}$

Quotient Law: In this law just subtract the exponents of the numerator and denominator having the same bases.

In symbol, $\dfrac{x^m}{x^n} = x^{m-n}$ where m>n

Example: 1.) $\dfrac{x^8}{x^6} = x^{8-6} = x^2$

2.) $\dfrac{n^7}{n^6} = n^{7-6} = n^1$ (or simply n)

   c.   *Law of Negative Exponents:* This law will be applied if the product or quotient of exponent happens to be negative. Reciprocal of the base will be used to make the exponent positive since expression with negative exponents is not accepted as final answer.

In symbol, $\dfrac{x^m}{x^n} = x^{m-n}$ where m<n

Example:
1.) $\dfrac{x^6}{x^8} = x^{6-8} = x^{-2} = \dfrac{1}{x^2}$

2.) $\dfrac{n^6}{n^7} = n^{6-7} = n^{-1} = \dfrac{1}{n}$

    d.   *Law of Zero Exponents*: Law applied when an exponent obtained is zero as a result of multiplying or dividing algebraic expressions, it should always become 1.

In symbol, $\dfrac{x^m}{x^n} = x^{m-n}$ where m=n

Example: 1.) $\dfrac{x^5}{x^5} = x^{5-5} = x^0 = 1$

2.) $\dfrac{n^9}{n^9} = n^{9-9} = n^0 = 1$

Application: Combination of the different laws

1. $(4x^5y^{-3})^2 = 16x^{10}y^{-6}$ or $\dfrac{16x^{10}}{y^6}$

2. $(12m^2n^2)^0 = 1$

3. $\dfrac{-36a^5b^8}{9a^7b^5} = -4a^{-2}b^3$ or $\dfrac{-4b^3}{a^2}$

4. $\dfrac{-45x^7y^3z^9}{-15x^2y^3z^{10}} = 3x^5y^0z^{-1} = \dfrac{3x^5}{z}$

5. $2(c^2d^3)^{-1} = \dfrac{2}{c^2d^3}$

Practice Set:

1. $\dfrac{(3x)^4}{3x^4}$

2. $\dfrac{(x^5)^3}{(x^4)^4}$

3. $\dfrac{(-2a)^3}{(a^2)^3}$

4. $\dfrac{(15a^4b)^2}{(15ab^2)^3}$

5. $\dfrac{(3x^4)(2x^5)}{(6x^3)^2}$

Chapter 10: Solving Algebra Problems

As mentioned on the introduction, there are variety of worded problems that maybe encountered wherein Algebra concepts is really necessary to be applied before one can solve a particular item. There are times only basic arithmetic is needed, though the problem looks so complex. But most of the time you have to use combinations of two or more of the previous lessons discussed.

Hence, on top of all the lessons discussed it is very important that you'll be able to comprehend, analyze the given situation and decide on what is the most appropriate approach that you are going to use. This simply require to plan on how to attack the problem per se and think of the best and shortest possible way to solve it.

The 3R's and the ESP in Solving Word Problems

Read the problem thoroughly

Represent the unknown using variables

Relate the unknown and the values given in the problem

Equate: Form an equation using the given facts

Solve the equation

Prove your answer/ Do the checking

Number Problems

Among all the word problems, it is considered as the easiest to translate into equations since the relationships among the numbers are directly stated.

Example1.) One number is two more than thrice another. Their sum is 30. What are the numbers?

READ: *By reading the problem thoroughly, two things have been mentioned about the numbers: First, their sizes: one of them is two more than thrice the other and second, their sum must be 30.*

We can use one of these sentences to represent the number and the other to form an equation.

REPRESENT: If we represent the numbers using the first sentence, we have

Let n = the first number

3n + 2 = the other number

RELATE: The relationship between the numbers n and 3n + 2, and 30 gives an equation

EQUATE: n + (3n + 2) = 30

SOLVE:
n + 3n + 2 = 30	Remove parenthesis
4n + 2 = 30	Combine similar terms
4n = 30 − 2	By SPE or transposition
4n = 28	Simplify
n = 28/4	By MPE or transposition
n = 7	

Therefore, the first number is 7 and the other number is 3n + 2 = 3(7) + 2 = 23.

PROVE: a. their sum is 30: 7 + 23 = 30

b. 23 is 2 more than thrice 7:

$$3(7) + 2 = 21 + 2 = 23$$

The answer satisfies both conditions of the problem.

2.) The sum of two numbers is 29 and their difference is 5. Find the numbers

READ: *By reading the problem thoroughly, two things have been mentioned about the numbers: First, their sum: should be 29 and second, their difference is 5.*

We can use one of these sentences to represent the number and the other to form an equation.

REPRESENT: If we represent the numbers using the first sentence, we have

Let n = one of the numbers

RELATE: 29 - n = the other number

The relationship between the numbers n and 29-n, and 5 gives an equation

EQUATE: n - (29 - n) = 5

SOLVE: n - (29 - n) = 5

$$n - 29 + n = 5 \quad \text{Remove parenthesis}$$
$$2n - 29 = 5 \quad \text{Combine similar terms}$$
$$2n = 5 + 29 \quad \text{By APE or transposition}$$
$$2n = 34 \quad \text{Simplify}$$
$$n = 34/2 \quad \text{By MPE or transposition}$$
$$n = 17$$

Therefore, the first number is 17 and the other number is 29 - n = 29 – 17 = 12.

PROVE: a. their sum is 29: 17 + 12 = 29

b. Their difference is 5: 17 – 12 = 5

The answer satisfies both conditions of the problem.

3.) The smaller of two numbers is thrice the larger. The larger number is eight more than the smaller one. Find the numbers.

READ: *The two facts about the numbers: First, the smaller of two numbers is thrice the larger and second, the larger is eight more than the smaller*

We can use the first sentence to represent the unknown numbers:

REPRESENT: If we represent the numbers using the first sentence, we have

Let n = the larger of the two numbers

RELATE: 3n = the smaller number

EQUATE: Using the second sentence, since the larger is eight more than the smaller (3n), then n = 3n + 8

SOLVE: n = 3n + 8

$\quad$ 3n + 8 = n $\qquad$ Using Symmetric Property
$\quad$ 3n − n = ⁻8 $\qquad$ By SPE or transposition
$\quad$ 2n = ⁻8 $\qquad$ Simplify
$\quad$ n = ⁻8/2 $\qquad$ By MPE or transposition
$\quad$ n = ⁻4

Therefore, the larger number is ⁻4 and the other number is 3n = 3(⁻4) = ⁻12.

PROVE: a. the smaller ⁻12 = 3 (⁻4)
$\qquad\qquad\qquad\qquad$ ⁻12 = ⁻12
$\qquad\quad$ b. the larger ⁻4 = ⁻12 + 8
$\qquad\qquad\qquad\qquad$ ⁻4 = ⁻4

The answer satisfies both conditions of the problem.

4.) The sum of two even numbers is 30. The larger number is twelve more than one half the smaller number. Find the numbers.

READ: *The problem provides three facts: First, the two numbers are even second, the sum is 30 and third, the larger number is 12 more than one half the smaller number.*

We can use the first sentence to represent the unknown numbers:

REPRESENT: We can use any of the sentences above to represent the unknowns and form an equation. Using the second condition:

Let n = the smaller number

RELATE: 30 - n = the larger number

EQUATE: Using the third condition,

$$30 - n = \tfrac{1}{2}n + 12$$

SOLVE: $30 - n = \tfrac{1}{2}n + 12$

$2(30 - n) = (\tfrac{1}{2}n + 12)2$	Using MPE
$60 - 2n = n + 24$	Distributive Property
$60 - 24 = n + 2n$	Using transposition
$36 = 3n$	Simplify
$\tfrac{36}{3} = n$	Use MPE or transposition
$12 = n$	

Therefore, the smaller number is 12 and the other number is 30 - 12 = 18.

PROVE: a. the two numbers are even: 12 and 18

b. their sum is 30: 12 + 18 = 30

c. the larger 18 = $\tfrac{1}{2}$(12) + 12 = 6 + 12 = 18

The answer satisfies all the conditions.

Practice Set:

1. Thirty-eight is equal to thrice a number increased by 5. Find the number.

2. If seven times a number is added to 33, the result is 75. Find the number.

3. One number is four more than three times another. Find the numbers if their sum is 60.
4. The difference between six times a number and twelve equals the sum of the number and thirteen. Find the number.
5. Three less than five times a certain number is equal to sixty-seven more than one-third of the number. Find the number.

Odd, Even and Consecutive Integers Problem

The word consecutive means following an exact order without interruption. Therefore, consecutive integers are integers which follow an order without interruption. Note that each consecutive integer exceeds the integer preceding it by 1.

Example1.) The sum of three consecutive integers is 90. Find the integers.

Representation:

Let x = first integer

$X + 1$ = second consecutive integer

$X + 2$ = third consecutive integer

Equation:

X + (x + 1) +(x + 2) = 90

Solution: X + (x + 1) +(x + 2) = 90

X + x + 1 + x + 2 = 90	Removing parenthesis
3x + 3 = 90	Combine similar terms
3x = 90 – 3	By SPE/transposition
3x = 87	Simplify
X = 87/3	By MPE/transposition

X = 29 is the first consecutive integer

Therefore, the three consecutive integers are 29, x + 1 = 29 + 1 = 30, x + 2 = 29 + 2 = 31.

Checking:

$$X + (x + 1) + (x + 2) = 90$$
$$29 + (29 + 1) + (29 + 2) = 90$$
$$29 + 30 + 31 = 90$$
$$90 = 90$$

2.) Find three consecutive even numbers whose sum is 108.

Representation:

Let x = first even integer

X + 2 = second consecutive even integer

X + 4 = third consecutive even integer

Equation:

$$X + (x + 2) + (x + 4) = 108$$

Solution: X + (x + 2) +(x + 4) = 108

X+ x + 2 + x + 4= 108   Removing parenthesis

3x + 6 = 108   Combine similar terms

3x = 108 − 6   By SPE/transposition

3x = 102   Simplify

X = 102/3   By MPE/transposition

X = 34 is the first consecutive integer

Therefore, the three consecutive even integers are 34, x + 2 = 34 + 2 = 36, x + 4 = 34 + 4 = 38.

Checking:

X + (x + 2) + (x + 4) = 108

34 + (34 + 2) + (34 + 4) = 108

34 + 36 + 38 = 108

108 = 108

3.) Find three consecutive odd integers whose sum is 57.

Representation:

Let x = first odd integer

X + 2 = second consecutive odd integer

X + 4 = third consecutive odd integer

Equation:

X + (x + 2) +(x + 4) = 57

Solution: X + (x + 2) +(x + 4) = 57

X + x + 2 + x + 4 = 57	Removing parenthesis
3x + 6 = 57	Combine similar terms
3x = 57 – 6	By SPE/transposition
3x = 51	Simplify
X = 51/3	By MPE/transposition

X = 17 is the first consecutive integer

Therefore, the three consecutive odd integers are 17, x + 2 = 17 + 2 = 19, x + 4 = 17 + 4 = 21.

Checking: X + (x + 2) + (x + 4) = 57

17 + (17 + 2) + (17 + 4) = 57

17 + 19 + 21 = 57

57 = 57

4.) Find two consecutive integers such that if we triple the first and double the second the sum is 77.

Representation:

Let x = first integer

X + 1 = second integer

Equation:

$3X + 2(x + 1) = 77$

Solution: $3X + 2(x + 1) = 77$

$3X + 2x + 2 = 77$	Distributive Property
$5x + 2 = 77$	Combine similar terms
$5x = 77 - 2$	By SPE/transposition
$5x = 75$	Simplify
$X = 75/5$	By MPE/transposition

$X = 15$ is the first integer

Therefore, the two integers are 1, x+1=15+1= 16.

Checking:

$$3X + 2(x + 1) = 77$$
$$3(15) + 2(15 + 1) = 77$$
$$45 + 2(16) = 77$$
$$45 + 32 = 77$$
$$77 = 77$$

Practice Set:

1.) Find three consecutive integers whose sum is -156.

2.) The sum of two consecutive odd integers is 188. Find the two integers.

3.) Find two consecutive even integers whose sum is 174.

4.) Find three consecutive integers such that the sum of the second and the third is twenty-four more than one-half the smallest.

5.) Find three consecutive even integers such that one-fourth the sum of the first and third is equal to 22 less than the second.

Digit Problems

Each digit differs in value from every other digit, thus in finding the value of a number, the position of each digit must be considered.

Example: 1.) The units digit in a two-digit number is one more than twice the tens digit. Find the number if the sum of the digits is 7.

Using 2 variables

Representation:
  Representation:

  Let t = tens digit   Let t = tens digit

2t + 1 = units digit   u = unit digit

Equation: t + (2t + 1) = 7   **Equations:**

Solution:

t + u = 7

t + (2t + 1) = 7   u = 2t + 1

t + 2t + 1 = 7   **Solution:** *by substitution*

3t + 1 = 7   t + u = 7

3t = 7 − 1   t + (2t + 1) = 7

3t = 6   t + 2t + 1 = 7

t = 6/3   3t + 1 = 7

t = 2   3t = 7 − 1

                                     3t = 6

                                     t = 6/3

                                     t = 2

Therefore, the tens digit is 2 and the units digit is 2t + 1 = 2(2) + 1 = 4 + 1 = 5

Checking:

The tens digit: 2

The units digit: 5 = 2(2) + 1

Their sum: 2 + 5 = 7

2.) The sum of the digits of a two-digit number is 9. The number is twelve times the tens digit. Find the number.

Representation:

Let u= units digit

9 - u = tens digit

Equation:

$10(9-u)+u=12(9-u)$

Solution:

$10(9-u)+u=12(9-u)$

$90-10u+u=108-12u$

$12u-10u+u=108-90$

$3u = 18$

$u = 18/3$

$u = 6$ is the units digit

$9 - u = 9 - 6 = 3$ is the tens digit

Using 2 variables

Representation:

Let u= units digit

t = tens digit

Equation:

$t+u=9 \longrightarrow t = 9-u$

$10t + u = 12t$

by substitution

$10t + u = 12t$

$10(9-u)+u=12(9-u)$

$90-10u+u=108-12u$

$12u-10u+u=108-90$

$3u = 18$

$u = 18/3$

$u = 6$

Therefore, the tens digit is 3 and the units digit is 6

Checking:

Their sum: 6 + 3 = 9

The tens digit: 3

The number: 36 = 12 (3)

3.) The sum of the digits of a two-digit number is 6. The number with the digits interchanged is three times the tens digit of the original number. Find the original number.

Representation:

Let u = units digit

6 − u = tens digit

10(6−u) + u = the original number

10u + 6 − u = the reversed number

Equation:

10u + 6 − u = 3(6 − u)

Representation:

u = units digit

t = tens digit

10t + u

10u + t

Equation:

t + u = 6 ⟶ t = 6 − u

10u + t = 3t

Solution:

9u + 6 = 18 – 3u
9u + 3u = 18 – 6
12u = 12
u = 1 units digit

Solution:

9u + 6 = 18 – 3u
9u + 3u = 18 – 6
12u = 12
u = 1

by substitution

10u + t = 3t
10u + 6 – u = 3(6 - u)
6 – u = 6 – 1 = 5 tens digit

Therefore, the original number is 51 and the reversed number is 15.

Checking:

Their sum: 1 + 5 = 6

The reversed number 15 = 3(5)

4.) The tens digit of a three-digit number is 0. The sum of the other two digits is 6. Interchanging the units and the hundreds digits decreases the number by 396. Find the original number.

Representation:

Let u = units digit

t = 0

6 − u = hundreds digit

100 (6−u) + 10(0) + u original number

100u + 10(0) + 6−u reversed number

Solution:

$100u + 10(0) + 6 - u = 100(6-u) + 10(0) + u - 396$

$100u + 6 - u = 600 - 100u + u - 396$

$100u - u + 100u - u = 600 - 396 - 6$

$198u = 198$

$u = \dfrac{198}{198}$

$u = 1$

Representation:

Let u= units digit

t = tens digit = 0

h= hundreds digit 6-u

100h+10t+u

100u+10t +h

Equations: $t = 0$

$u + h = 6$

$100u + 10t + h = 100h + 10t + u - 396$

Solution: *by substitution*

$100u + 10t + h = 100h + 10t + u - 396$

$100u + 10(0) + 6 - u = 100(6-u) + 10(0) + u - 396$

$100u + 6 - u = 600 - 100u + u - 396$

$100u - u + 100u - u = 600 - 396 - 6$

$198u = 198$

$u = \dfrac{198}{198}$

$u = 1$

Therefore, the original number is

$100(6-u) + 10t = 100(6 - 1) + 10(0) + 1$

$= 100(5) + 0 + 1$

$= 500 + 1$

$= 501$

Checking:

The sum of the digits is $6 = 5 + 0 + 1$

The reversed number $105 = 501 - 396$

$105 = 105$

Practice Set:

1. The sum of the digits of a two-digit number is 12. The value of the number is equal to eleven times the tens digit. Find the number.
2. The units digit of a two-digit number exceeds thrice the tens digit by 1. The sum of the digits is 9. Find the number.
3. The sum of the digits of a two-digit number is 15. One-third of the number diminished by 12 is 17. Find the number.
4. The tens digit of a two-digit number is 3 less than the units digit. The number is four times the sum of the digits. Find the number.
5. The sum of the digits of a two-digit number is 8. The number with the digits interchanged is fourteen more than eight times the units digit of the original number. Find the original number.

Age Problems

In dealing with problems about age it is important to take note that anybody's ages change at the same rate. As one goes old the other does too.

Example1.) Glenn is now 20 years older than his son. In ten years, he will be twice as old as his son's age. What are their present ages?

Representation: Using one variable

	Present Age	Age in 10 yrs
Son	x	X + 10
Glenn	X + 20	(x + 20) + 10

Equation:

X + 20 + 10 = 2(x + 10)

Solution:

X + 20 + 10 = 2(x + 10)

X + 30 = 2x + 20

-20 + 30 = 2x − x

10 = x   son's age now

x + 20 = 10 + 20 = 30 Glenn's age now   **Or**

Representations: *Using 2 variables*

x = son's age now

y = Glenn's age now

x + 10 = son's age in 10 years

y + 10 = Glenn's age after 10 years

Equations:

y = x + 20

y + 10 = 2(x + 10)

Solution: *by substitution*

y + 10 = 2(x + 10)

(x + 20) + 10 = 2x + 20

x + 30 = 2x + 20

30 − 20 = 2x − x

10 = x   son's age now

In 10 years, son will be 20 years old and Glenn will be 40 years old.

Checking:

Glenn is now 30 years old, which is equal to 20 years older than his son who is 10 years old.

After 10 years, he will become 40 years old = twice his son's age then, which is 20 years old.

2.) The sum of Nini and Nina's age is 60. Nine years ago, Nini was twice as old as Nina then. How old is Nina?

Representation:

	Now	**9 years ago**
Nina	x	x - 9
Nini	60 - x	(60 − x) − 9 **or** 51 - x

Equation:

$51 - x = 2(x - 9)$

Solution:

$51 - x = 2(x - 9)$

$51 - x = 2x - 18$

$51 + 18 = 2x + x$

$69 = 3x$

$69/3 = x$

$23 = x$ \quad Nina's present age

$60 - x = 60 - 23 = 37$ \quad Nini's present age \quad **or**

Representations: *Using 2 variables*

Let x = Nina's present age

y = Nini's present age

x − 9 Nina's age 9 years ago

y − 9 Nini's age 9 years ago

Equations:

$x + y = 60 \longrightarrow y = 60 - x$

$y - 9 = 2(x - 9)$

Solution: *by substitution*

$y - 9 = 2(x - 9)$

$60 - x - 9 = 2x - 18$

$51 + 18 = 2x + x$

$69 = 3x$

$69/3 = x$

$23 = x$ Nina's age now

$y = 60 - x$

$y = 60 - 23$

$y = 37$ Nini's age now

Checking:

The sum of their ages is 60 = 23 + 37

Nine years ago, Nini is 28 years old which is equal to twice Nina's age (14 years old) then.

3.) Aimee is twice as old as Rica while Jay is 24 years younger than Aimee. If half of Aimee's age six years ago was three less than $\frac{1}{2}$ the sum of Rica's age in 4 years and Joy's present age. Find their present ages.

Representation:

	Present Age	Age 6 years ago	Age 4 years hence
Rica	x	X - 6	X + 4
Aimee	2x	2x - 6	2x + 4
Joy	2x - 24	(2x−24) -6 or 2x - 30	2x − 24 + 4 or 2x - 20

Equation:

$\frac{1}{2}(2x - 6) = \frac{1}{2}[(x + 4) + (2x - 24)] - 3$

2$\{\frac{1}{2}(2x - 6) = \frac{1}{2}[(x + 4) + (2x - 24)] - 3\}$**2**

$2x - 6 = x + 4 + 2x - 24 - 6$

$-4 \cancel{-6} + 24 + \cancel{6} = x - \cancel{2x} + \cancel{2x}$

$20 = x$ Rica's present age

$2x = 2(20) = 40$ Aimee's age now

$2x - 24 = 2(20) - 24 = 40 - 24 = 16$ Joy's age now

or

Representations: *Using 3 variables*

	Present Age	Age 6 years ago	Age 4 years hence
Rica	x	x - 6	x + 4
Aimee	y = 2x	y - 6	y + 4
Joy	z = y − 24 or 2x - 24	z - 6	z + 4

Equations:

$\frac{1}{2}(y - 6) = \frac{1}{2}[(x + 4) + z] - 3$

Solution: *by substitution*

$\frac{1}{2}(2x - 6) = \frac{1}{2}[(x + 4) + (2x - 24)] - 3$

$2\{\frac{1}{2}(2x - 6) = \frac{1}{2}[(x + 4) + (2x - 24)] - 3\}2$

$2x - 6 = x + 4 + 2x - 24 - 6$

$-4 - 6 + 24 + 6 = x - 2x + 2x$

$20 = x$ Rica's present age

$y = 2x = 2(20) = 40$ Aimee's age now

$z = 2x - 24 = 2(20) - 24 = 16$

Checking:

Aimee is twice as old as Rica 40 = 2 (20)

Joy is 24 years younger than Aimee 16 = 40-24

Half of Aimee's age 6 years ago is equal to three less than one half of the sum of Rica's age in 4 years (24) and Joy's present age, (16) were added. That is:

$$\frac{1}{2}(34) = \frac{1}{2}(24 + 16) - 3$$

$$17 = \frac{1}{2}(40) - 3$$

$$17 = 17$$

Practice Set

1. Edna is six years older than his brother Edgar. In two years, Edna will be 20 more than one-thirds of Edgar's age. How old are they now?

2. A father is 8 times as old as his daughter and four times as old as his son. If the son is four years older than the daughter, how old is each?

3. Mira is four times as old as Yvonne. Five years ago, Yvonne was 30 years younger than Mira. Find their ages now.

4. When Melvin and Mercy got married, Mercy was 2 years younger than Melvin. How old is Melvin during the wedding day, if 16 years ago he was twice as old as Mercy then?

5. When Evan is twelve years old, his father was forty years old. How many years should pass for Evan to be eight years older than $\frac{1}{4}$ as old as his father?

Geometry Problems

Finding the measures of the missing part/s of a geometric figure can also be obtained using linear equations.

Example1.) A rectangle is 3 cm longer than its width. If its perimeter is 54 cm, find the dimensions of the rectangle.

Representation:

Let w = width

  L = length = (w + 3)

Note the formula in finding the perimeter of rectangle: P = 2l +2w

Equation:
  w + 3

  l = w + 3

  54 = 2l + 2w   w

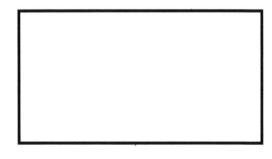

Solution: *by substitution*

54 = 2l + 2w
54 = 2(w + 3) + 2w
54 = 2w + 6 + 2w
54 − 6 = 4w
48 = 4w
48/4 = w
12 = w

l = w + 3
l = 12 + 3
l = 15

Checking:

l = w + 3
15 = 12 + 3
15 = 15

or

54 = 2(15) + 2(12)
54 = 30 + 24
54 = 54

2. Two sides of a triangle are 2 cm longer and 6 cm longer than the shortest side. Find the lengths of the sides of the triangle when its perimeter is 62 cm.

Representation:

Let x = length of the shortest side
x + 2 = length of the second side
x + 6 = length of the longest side

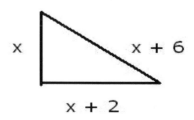

Equation:

P = a + b + c

62 = x + (x + 2) + (x + 6)

62 = 3x + 8

62 − 8 = 3x

54 = 3x

54/3 = x

18 = x   *length of the shortest side*

X + 2 = 18 + 2 = **20**   *length of the second side*

X + 6 = 18 + 6 = **24**   *length of the longest side*

Checking:

P = x + x + 2 + x + 6

62 = 18 + 18 + 2 + 18 + 6

62 = 62

Practice Set

1. In triangle XYZ, side XY is 2 cm shorter than side XZ, while side YZ is 1 cm longer than XZ. If the perimeter is 62 cm, find the lengths of the three sides.

2. The length of a rectangle is 3 m less than twice its width. The perimeter of the rectangle is 90 m. What are the dimensions of the rectangle?

3. The lengths in meters of the sides of a triangle are consecutive even integers. The perimeter is 210 cm. What is the length of the sides?

4. The width of a rectangle is 4 cm shorter than its length. If the perimeter of the rectangle is 96 cm, what are its dimensions?

5. The lengths of the sides of a triangle are consecutive odd integers, if its perimeter is 171 cm. How long is each side?

Distance/Motion Problems

When an object moves without changing its speed or rate, it is said to be in uniform motion. There are three possible motion problems; motion in the same direction, motion in opposite direction and motion involving round-trips.

Example1.) There are two public utility buses, bus A heads north on the expressway at 45kph.

After 12 minutes, bus B follows at a steady rate of 54kph. How long does it take bus B to overtake bus A?

Representation:

Let x = Bus B's time

X + 0.20 = Bus A's time

(Note:12mins. = 1/5 of an hr or equivalent to 0.20)

	Rate x	Time =	Distance
Bus A	45	X + 0.20	45(x+0.20)
Bus B	54	x	54x

Equation:

45(x + 0.20) = 54x

Solution:

45x + 9 = 54x

9 = 54x - 45x

9 = 9x

9/9 = x

1 = x

Therefore, bus B overtakes bus A in 1 hour.

2.) Mr. Roberts and Mr. Rogers agreed to meet on the highway connecting their hometowns. Mr. Roberts drives at 45 kph and Mr. Rogers at 35kph. They leave their homes which are 120 km apart at the same time. In how many hours will they meet?

Representation:

Let t = the number of hours before they meet

	Rate x	Time =	Distance
Mr. Roberts	45	t	45t
Mr. Rogers	35	t	35t

Equation:

45t + 35t = 120

Solution:

45t + 35t = 120

80t = 120

t = 120/80 or 3/2 or $1\frac{1}{2}$ hours

Checking:

After $1\frac{1}{2}$ hours Mr. Roberts gone 3/2 x 45 = 67.5

After $1\frac{1}{2}$ hours Mr. Rogers gone 3/2 x 35 = <u>52.5</u>

The total distance they travel is equal to   120.0

3.) While waiting for a train, Anne takes a bus ride at 60kph to a certain point and then walks back leisurely at 10kph. The bus ride takes 15 minutes less than her walking. How far does she walk and for how long?

Representation:

Let t = the walking time

10t = the walking distance

Note that 15 min. = 0.25 hr.

	Rate x	Time =	Distance
Walking	10	t	10t
Riding	60	t – 0.25	60(t - 0.25)

In going back and forth the distance are equal.

Equation:

$60(t - 0.25) = 10t$

$60t - 15 = 10t$

$60t - 10t = 15$

$50t = 15$

$t = 15/50$ or $3/10$ or 0.3 hr

To compute the distance $10t = 10(0.3) = 3$km.

Checking:

Anne walks $10(0.3)$ = a distance of 3km and has ridden $60(0.3 - 0.25)$ = which is also 3km.

Practice Set

1. Two airplanes leave the airport at 4p.m., one travelling east at 850 kph and the other west at 750 kph. At what time will they be 4000 km apart?
2. Mikee can bike 5 kph faster than Marianne. At noon, each girl leaves her house and travels towards each other. If the distance

between their houses is 15 km and they meet in 30 minutes what is each girl's rate?

3. A freight train leaves the station at a rate of 60kph. Three hours later, an express train with a rate of 80kph leaves the station travelling the same direction. How long will it take the express train overtake the freight train?

4. In a track race Lily is 50n ft. in front of Edna in 10 sec. How fast can Lily run if Edna runs 20 ft /sec?

5. Two cars start at the same point at the same time and travel in opposite directions. The faster car travels 12mph than the other. If after 6 hrs, the cars are 648 km apart, what is the rate of speed of each car?

Conclusion

There is no other subject that can offer preciseness and accuracy of answers other than mathematics, most specifically Algebra.

As a problem solver, you are given variety of options as to the method you are going to use to solve a particular problem and how you wanted to present your answer. That is contingent on which of these you decide to go with. Feel free to make use of visual learning aids like diagrams and tables. This is to clearly see the scenario being presented. You may also be required to work backwards sometimes when the situation demands.

The most important trait you can have is to simply be smart. Mathematics is a symbolic interpretation of problems in the real world. Being smart really helps you understand this. The easier it is for you to comprehend things, the easier learning mathematics will be for you.

This e-book has not even brushed the surface of mathematics. We have not talked about trigonometry, calculus, or physics, which are all about ten times harder than this. Mathematics continues to amaze me because of its uncanny resemblance to driving a car. No matter what route you decide to take, you will always arrive at the same destination.

As you develop a more solid understanding of algebra and mathematics in general, you will attempt to solve problems using the most efficient, intelligent solution way, whilst staying in accordance with the basic laws and rules of the underlying discipline.

As we mentioned at the beginning of this guide, practice makes perfect. The more you practice, the less confusing algebra will be, including story problems. Eventually you will be doing algebra problems in your head. The more you practice, the faster of a problem solver you will become. The more problems solved, the more you discover how easy it is to deal with the subject. Once you start to learn algebra and grasp the basic concepts, you will develop a strong feeling of confidence when doing algebra, and you may even come to love the subject matter. Once you master the most difficult subject in mathematics, quantum physics, then you will not have to worry about finding other mathematics subjects difficult anymore.

And you can now enjoy a worry-free study. This way you will not have to be worried about going to school and humiliating yourself in front of your peers and teacher because you will actually have a clue about what it is that you are doing.

Though there are still other types of problems like investment, mixture, work, miscellaneous etc. that you might encounter, you just have to apply varied approach but stick on the same concepts. Always remember to focus on mastering the basic, fundamental concepts first. Everything in mathematics builds on this principle.

Recommended Resources

- HowExpert.com – Quick 'How To' Guides on All Topics from A to Z by Everyday Experts.
- HowExpert.com/free – Free HowExpert Email Newsletter.
- HowExpert.com/books – HowExpert Books
- HowExpert.com/courses – HowExpert Courses
- HowExpert.com/clothing – HowExpert Clothing
- HowExpert.com/membership – HowExpert Membership Site
- HowExpert.com/affiliates – HowExpert Affiliate Program
- HowExpert.com/writers – Write About Your #1 Passion/Knowledge/Expertise & Become a HowExpert Author.
- HowExpert.com/resources – Additional HowExpert Recommended Resources
- YouTube.com/HowExpert – Subscribe to HowExpert YouTube.
- Instagram.com/HowExpert – Follow HowExpert on Instagram.
- Facebook.com/HowExpert – Follow HowExpert on Facebook.

Printed in Poland
by Amazon Fulfillment
Poland Sp. z o.o., Wrocław